BODY SYSTEMS

The Nervous System

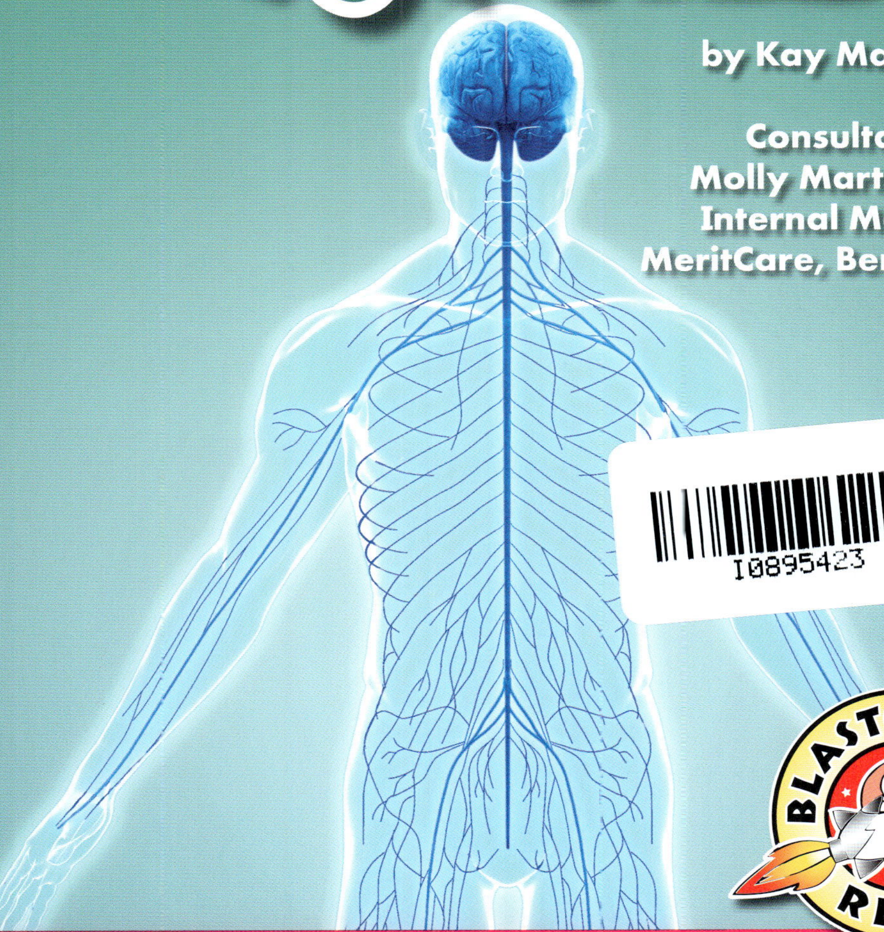

by Kay Manolis

Consultant:
Molly Martin, M.D.
Internal Medicine
MeritCare, Bemidji, MN

BLASTOFF! READERS 4

BELLWETHER MEDIA • MINNEAPOLIS, MN

Note to Librarians, Teachers, and Parents:

Blastoff! Readers are carefully developed by literacy experts and combine standards-based content with developmentally appropriate text.

Level 1 provides the most support through repetition of high-frequency words, light text, predictable sentence patterns, and strong visual support.

Level 2 offers early readers a bit more challenge through varied simple sentences, increased text load, and less repetition of high-frequency words.

Level 3 advances early-fluent readers toward fluency through increased text and concept load, less reliance on visuals, longer sentences, and more literary language.

Level 4 builds reading stamina by providing more text per page, increased use of punctuation, greater variation in sentence patterns, and increasingly challenging vocabulary.

Level 5 encourages children to move from "learning to read" to "reading to learn" by providing even more text, varied writing styles, and less familiar topics.

Whichever book is right for your reader, Blastoff! Readers are the perfect books to build confidence and encourage a love of reading that will last a lifetime!

This edition first published in 2016 by Bellwether Media, Inc.

Library of Congress Cataloging-in-Publication Data
Manolis, Kay.
Nervous system / by Kay Manolis.
p. cm. – (Blastoff! Readers: body systems)
Includes bibliographical references and index.
Summary: "Introductory text explains the functions and physical concepts of the nervous system with color photography and simple scientific diagrams. Intended for students in grades three through six"–Provided by publisher.
ISBN: 978-1-60014-245-1 (hardcover : alk. paper)
ISBN: 978-1-62617-472-6 (paperback : alk. paper)
1. Nervous system–Juvenile literature. I. Title.
QP361.5.M36 2009
612.8–dc22 2008032701

Printed in the United States of America, North Mankato, MN.

Contents

What Is the Nervous System?

What did you do today? Did you jump or run? Did you hear or touch anything? Whatever you did today, your nervous system made it possible.

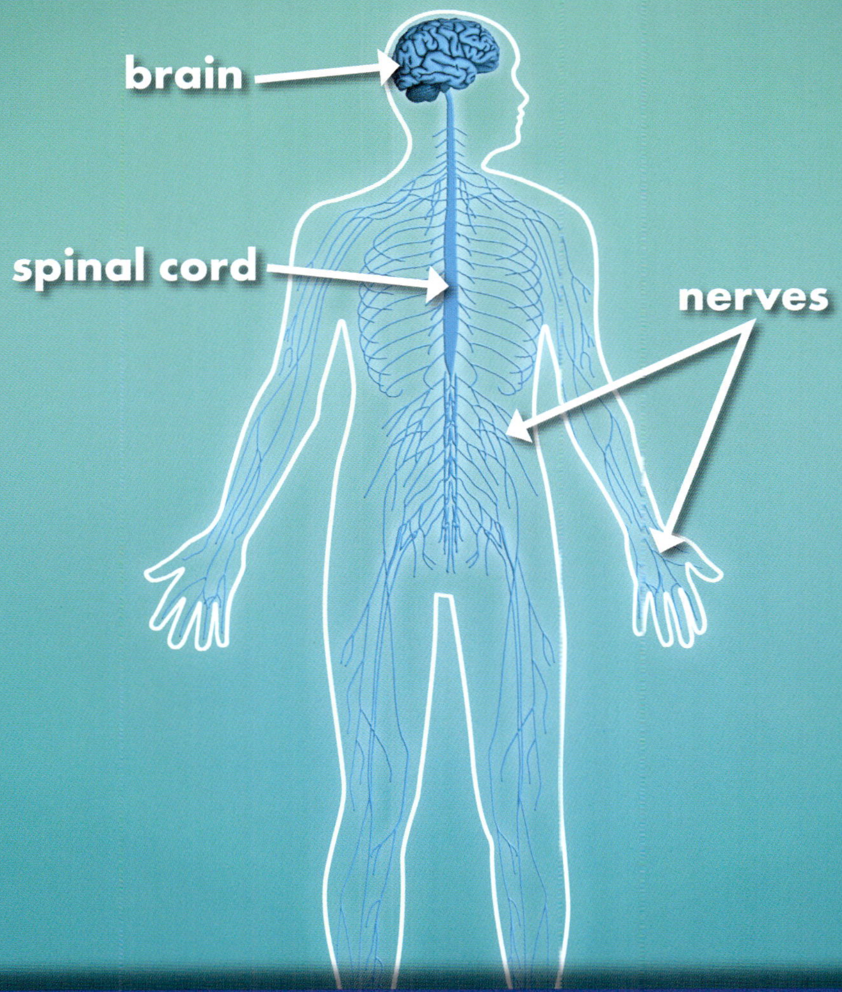

Your nervous system includes your **brain**, **spinal cord**, and **nerves**. It is the system that lets you experience the world around you. It also controls the other systems that work together to make your body function.

Parts of the Nervous System

The Brain

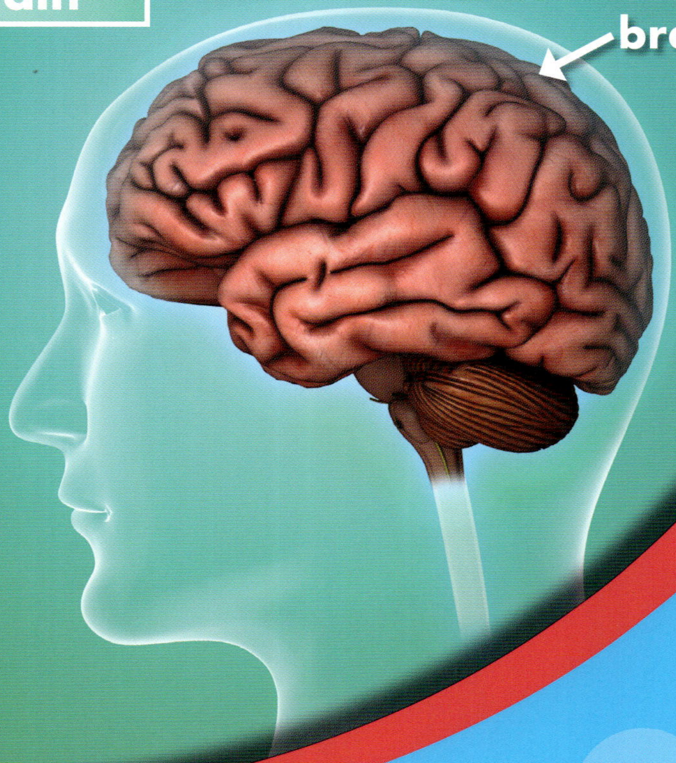

Your brain, like all parts of the nervous system, is filled with **cells** called **neurons**. Billions of neurons in the brain send messages to one another. These messages are responsible for your thoughts, memories, and feelings.

A neuron has several parts. It has a body with many short arms called **dendrites**. Dendrites receive messages from other neurons. Neurons also have a long tail called an **axon**. Axons send messages to other cells.

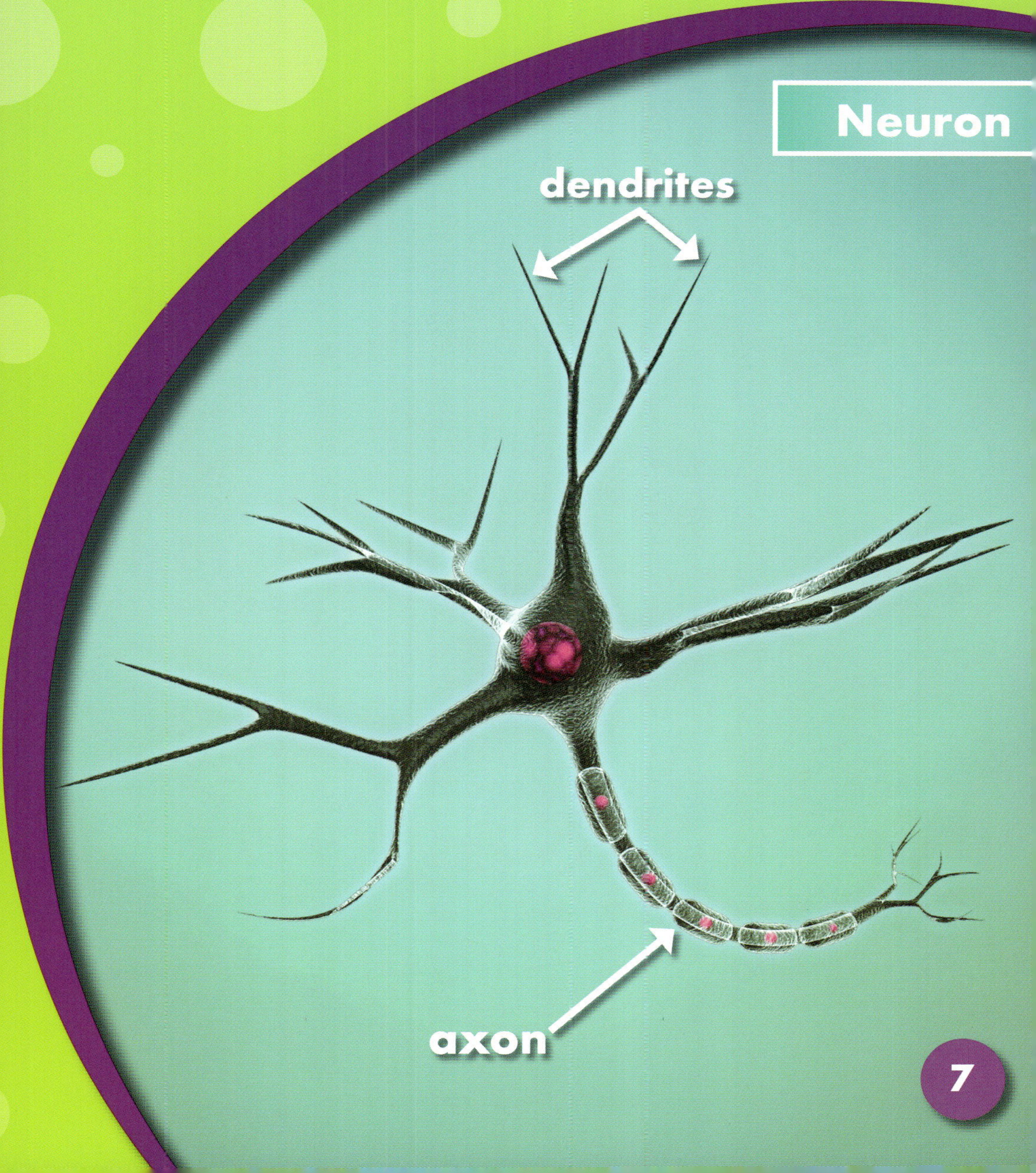

fun fact

An adult brain weighs less than 3 pounds (1.4 kilograms).

Neurons and other cells join together to form nerves. Nerves run through your body. They carry messages from your brain to the rest of your body. These messages tell your body how to move.

Nerves also carry messages from your body back to your brain. These messages help you experience the world.

The Spinal Cord

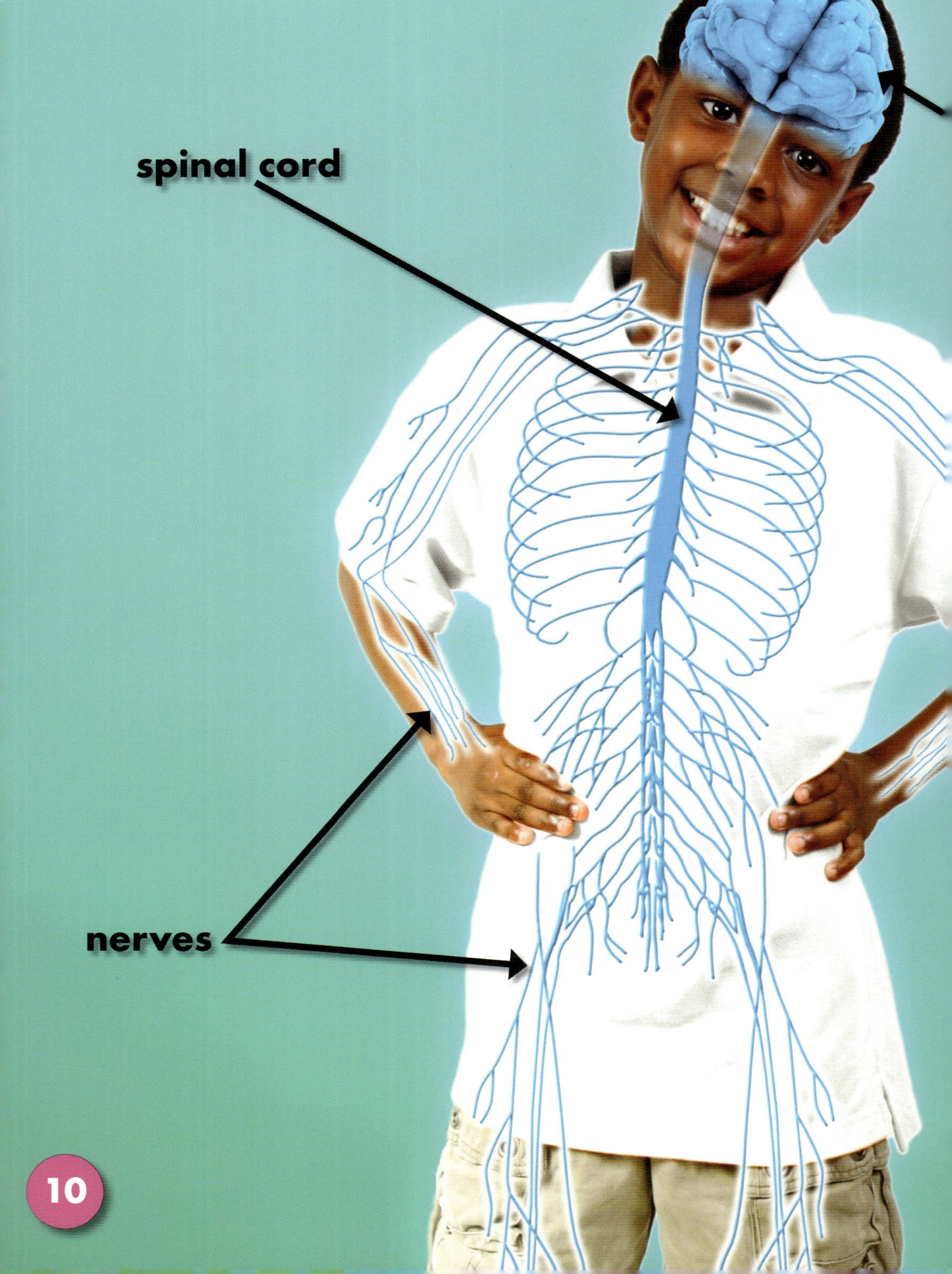

The spinal cord is a very thick bundle of nerves that connects your brain to the nerves of your body. Most messages enter and leave your brain through the spinal cord.

fun fact

If you laid out the nerves from an adult's body end to end, they would be about 47 miles (76 kilometers) long.

What Does the Nervous System Do?

The nervous system is responsible for making your body function. You are not aware of many of the messages that your brain sends out.

Messages from your brain tell your heart to beat. Messages to your throat tell it to swallow. Messages to your lungs tell them to breathe.

The nervous system also controls actions that you do think about. For example, an athlete wants to kick a ball. Her brain sends a message through nerves to her **muscles**. Her muscles move her leg to kick the ball.

fun fact

Nerve messages can zip from one nerve cell to another at speeds of 70 miles (113 kilometers) per hour or faster.

The nervous system allows your **senses** to collect information. Sight, touch, smell, taste, and hearing are your body's senses.

Nerves connect to **sensors** in your eyes, ears, nose, mouth, and skin. These sensors receive information from the world around you.

For example, sensors can pick up information about the sight, smell, taste, or feel of freshly baked dessert. Nerves carry the information from the sensors to your brain. Your brain makes sense of the information.

fun fact

Nerves make your eyelids blink if something comes close to your eyes. Blinking stops small objects from getting into your eyes.

Your nervous system lets you feel the warmth of a hug. It also helps to keep you safe.

If you touch something that is very hot, skin sensors send a message to your brain. Your nervous system causes you to quickly pull your hand away.

Taking Care of Your Nervous System

Your nervous system helps keep you safe, runs your body, and lets you experience the world.

You can help keep your nervous system safe. Eating healthy foods helps your brain work its best. Use a helmet when you do activities in which you could hurt your head. Protect your nervous system like it protects you!

Glossary

axon—the part of a neuron that sends information; this part is also called the nerve fiber.

brain—the organ of the body inside the skull; the brain is filled with nerve cells and controls the nervous system.

cells—the basic building blocks of living things

dendrite—a part of a neuron that takes in information

muscle—a body part that can squeeze to produce force or motion

nerves—bundles of neurons throughout your body that carry messages between your brain and other parts of your body

neurons—the most important kind of cells that make up the nervous system; neurons send and receive messages that control your body's functions and movements, and let you experience your environment.

senses—ways of getting information from the world through parts of the body; the body's senses are sight, touch, smell, taste, and hearing.

sensor—tiny parts in the body that help you touch, taste, see, smell, and hear; nerves deliver information from sensors to the brain.

spinal cord—a thick cord of nerves running down the middle of your back that carries information to the brain and all parts of the body; the spinal cord is protected by the bones of the spine.

To Learn More

AT THE LIBRARY

Riley, Joelle. *The Nervous System.* Minneapolis, Minn.: Lerner, 2004.

Simon, Seymour. *The Brain: Our Nervous System.* New York: HarperCollins, 2006.

Taylor-Butler, Christine. *The Nervous System.* New York: Children's Press, 2008.

ON THE WEB

Learning more about the nervous system is as easy as 1, 2, 3.

1. Go to www.factsurfer.com.
2. Enter "nervous system" into the search box.
3. Click the "Surf" button and you will see a list of related Web sites.

With factsurfer.com, finding more information is just a click away.

Index

The images in this book are reproduced through the courtesy of: Sebastian Kaulitzki, front cover; Mandy Godbehear, p. 4; Linda Clavel, diagrams, pp. 5, 10-11; Anatomical Design, p. 6; sgame, p. 7; Peter Weber, p. 8; Tim Hall / Getty Images, p. 9; Doug Schneider, pp. 10-11; Rhienna Cutler, p. 12; Jacek Chabraszewski, p. 13; John Giustina / Getty Images, pp. 14-15; Jacques Alexandre / age fotostock, p. 16; digitalskillet, p. 18; iofoto, p. 19; Tomasz Trojanowski, pp. 20.